W9-AMC-547

ABOUT THE BANK STREET READY-TO-READ SERIES

More than seventy-five years of educational research, innovative teaching, and quality publishing have earned The Bank Street College of Education its reputation as America's most trusted name in early childhood education.

Because no two children are exactly alike in their development, the Bank Street Ready-to-Read series is written on three levels to accommodate the individual stages of reading readiness of children ages three through eight.

○ *Level 1:* **GETTING READY TO READ (Pre-K–Grade 1)**
Level 1 books are perfect for reading aloud with children who are getting ready to read or just starting to read words or phrases. These books feature large type, repetition, and simple sentences.

○ *Level 2:* **READING TOGETHER (Grades 1–3)**
These books have slightly smaller type and longer sentences. They are ideal for children beginning to read by themselves who may need help.

○ *Level 3:* **I CAN READ IT MYSELF (Grades 2–3)**
These stories are just right for children who can read independently. They offer more complex and challenging stories and sentences.

All three levels of the Bank Street Ready-to-Read books make it easy to select the books most appropriate for your child's development and enable him or her to grow with the series step by step. The levels purposely overlap to reinforce skills and further encourage reading.

We feel that making reading fun is the single most important thing anyone can do to help children become good readers. We hope you will become part of Bank Street's long tradition of learning through sharing.

The Bank Street College of Education

For Fred, who found me
— T.S.

To Mrs. Naomi Younger, my kindergarten
teacher, who let me take
books home.
— D.B.

For a free color catalog describing Gareth Stevens' list of high-quality books and multimedia programs, call 1-800-542-2595 (USA) or 1-800-461-9120 (Canada). Gareth Stevens Publishing's Fax: (414) 225-0377.
See our catalog, too, on the World Wide Web: http://gsinc.com

Library of Congress Cataloging-in-Publication Data

Slater, Teddy.
    Animal hide-and-seek / by Teddy Slater; illustrated by Donna Braginetz.
      p. cm. -- (Bank Street ready-to-read)
      Summary: Examines how and why certain birds, insects, and mammals disappear by blending in with their backgrounds.
      ISBN 0-8368-1760-5 (lib. bdg.)
      1. Camouflage (Biology)--Juvenile literature. [1. Camouflage (Biology)
2. Animal defenses.] I. Braginetz, Donna, ill. II. Title. III. Series.
QL759.S58 1998
591.47'2--dc21   97-30141

This edition first published in 1998 by
Gareth Stevens Publishing
1555 North RiverCenter Drive, Suite 201
Milwaukee, Wisconsin 53212 USA

© 1997 by Byron Preiss Visual Publications, Inc. Text © 1997 by Bank Street College of Education. Illustrations © 1997 by Donna Braginetz and Byron Preiss Visual Publications, Inc.

Published by arrangement with Bantam Doubleday Dell Books for Young Readers, a division of Bantam Doubleday Dell Publishing Group, Inc., New York, New York. All rights reserved.

Bank Street Ready To Read ™ is a registered U.S. trademark of the Bank Street Group and Bantam Doubleday Dell Books For Young Readers, a division of Bantam Doubleday Dell Publishing Group, Inc.

Printed in Mexico

1 2 3 4 5 6 7 8 9 02 01 00 99 98

Bank Street Ready-to-Read™

ANIMAL HIDE-AND-SEEK

by Teddy Slater
Illustrated by Donna Braginetz

A Byron Preiss Book

Gareth Stevens Publishing
MILWAUKEE

INTRODUCTION

The animals in this book play
hide-and-seek all the time.
We'll tell you how they play.
See if you can figure out why.

BLENDING IN

Now you see it . . .

Now you don't!

The bittern is sometimes called
"the invisible bird."

It's not hard to see why.
At the first sign of danger,
this bird stands tall and stretches
its neck to blend in with the reeds.

What do you think it's hiding from?

You may not think this sea horse
is very well hidden.
But there's more here
than meets the eye.

Try to trace the outline of the
sea horse with your finger.

Now try again.

Surprise! This is not a sea horse.
It's a sea dragon!
And that green stuff isn't seaweed.
It's the sea dragon's leafy skin.

In winter the arctic hare almost
disappears in the bright white snow.
But so does the fox sneaking up on it.

When the snow melts each spring,
both animals shed their white fur.
They grow new brown coats
to match the bare earth.

Do you see a gaboon viper
curled up on the ground?

Look again.
There are two!

Most animals must stay still to hide.
But even when it's on the prowl,
the tiger's wavy stripes
blend in with the waving grass.

PROTECTING THE YOUNG

Baby animals can't run fast.
They're not strong enough to fight.
But they can hide
from hungry hunters.

The safest place for a baby zebra
is right beside its mother.
Their stripes blend together.
It's hard to tell where one begins . . .
and the other one ends.

Harp seals are born on
the pale, frozen ground.
Their thick white fur is
a perfect cover-up.

When they learn how to swim,
the seals take to the sea.
Then their fluffy white fur
turns dark and sleek.

This young deer's spots look like
patches of sun on the forest floor.
The spots won't fade till the fawn
is old enough to run from danger.

FLOWER POWER

The brightly colored flower mantis
looks like a jungle blossom.
Can you tell which part of the
flower is really a hungry hunter?

This butterfly couldn't!

The sea anemone's pretty petals
are really tentacles.
When a small fish swims near,
the tentacles close around it.
Then—PLOP!
The fish is in the anemone's mouth.

DRESSING UP

The decorator crab sticks pieces
of sponge, moss, and seaweed onto
the sharp spines that cover its body.

Every time the crab moves to a
different place, it takes off its
old costume and makes a new one.
For this clever crab,
every day is Halloween!

Crickets hunt in the dark of night.
They sleep in the light of day.
To avoid being eaten by birds or
bigger insects, the cricket makes a
special sleeping bag where it can hide.

Each morning the cricket rolls a
piece of leaf around its body.
It "sews" the edges together with
silk thread made by glands
            in its mouth . . . and disappears!

QUICK-CHANGE ARTISTS

Sometimes they are yellow.
Sometimes they are pink.
Crab spiders change color to match
the flowers they sit on.

27

How many different fish
can you see on these pages?
Four?
Guess again—there's only one!
A flounder can be striped one
minute . . . and spotted the next.
It can even look
like a checkerboard!

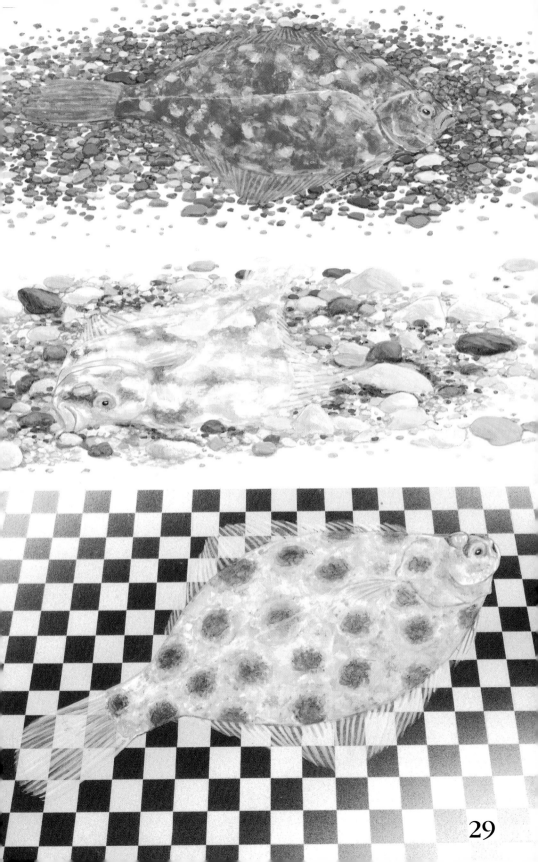

Chameleons also change color.
Six of these little lizards
are hiding in the woods.
Can you find them all?

Can you spot some other
colorful creatures?

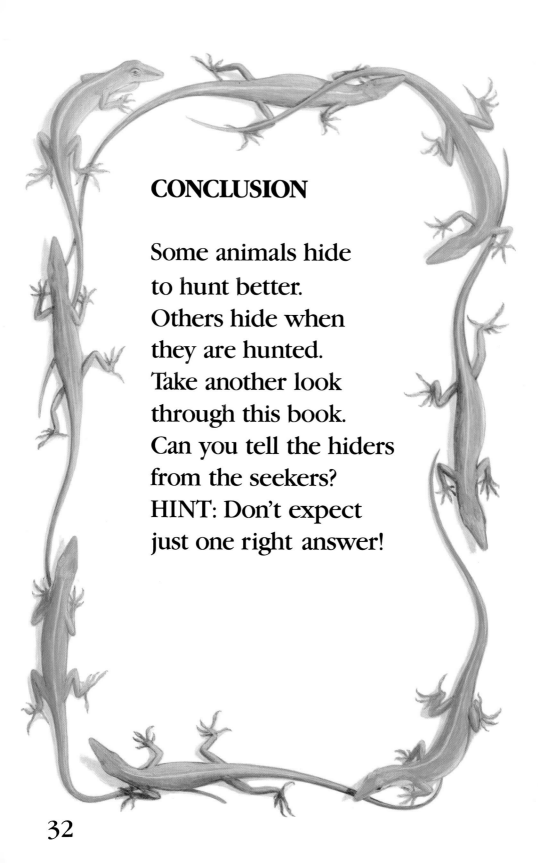

CONCLUSION

Some animals hide
to hunt better.
Others hide when
they are hunted.
Take another look
through this book.
Can you tell the hiders
from the seekers?
HINT: Don't expect
just one right answer!